BITCOIN

Concise Guide on Buying and Selling of Bitcoin

BY

PAUL L. MULLEN

COPYRIGHT

TABLE OF

CONTENTS

CHAPTER 1..4

INTRODUCTION 4

CHAPTER 2..6

WHAT IS BITCOIN?..6

CHAPTER 3..9

BITCOIN AND TRADITIONAL CURRENCY? 9

CHAPTER 4..11

HOW BITCOIN WORKS? 11

CHAPTER 5..15

LAWS FOR CREATING BITCOIN ACCOUNT 15

CHAPTER 6..20

HOW TO PURCHASE BITCOIN? 20

CHAPTER 7..24

MORE OF BITCOIN 24

THE END..25

CHAPTER 1

INTRODUCTION

Bitcoin is a type of legal tender in which encryption techniques are used to control the generation of units of currency and validate the transfer of funds, it operates independently of a central bank. Bitcoin can be used for online transactions between persons; it has become a most wanted commodity among speculators.

Bitcoin is a crypto currency, a form of electronic currency. It has no central bank or either a particular controller which also makes it a decentralized digital currency. Bitcoin doesn't need mediators while transferring it directly from peer-to-peer on the Bitcoin user network.

The Block chain is a public circulated ledger that verifies transaction made with bitcoin by network nodes through cryptography.

Bitcoin allows payment between two various users without passing through a central government. Bitcoin is also fashioned electronically and they aren't been printed.

Bitcoin is formed by computers all around the world using the software.

CHAPTER 2

WHAT IS BITCOIN?

Bitcoin is a sort of currency in which encryption techniques are used to normalize the generation of units of currency and verify the transfer of funds, operating autonomously of a central bank. Bitcoin can be used for online transactions between individuals; it has become a high commodity among speculators.

Bitcoin is a crypto currency, a form of electronic money. It has no central bank or either a single controller which also makes it a decentralized digital currency. Bitcoin doesn't need mediators while sending it directly from peer-to-peer on the Bitcoin user network.

The Block chain is an unrestricted circulated ledger which verifies transaction made with bitcoin by network nodes through cryptography.

Bitcoin allows payment to be sent between two users without going through a central government. Bitcoin is also created and managed electronically and aren't been printed.

Bitcoin is produced by computers all around the world using the software.

Origin of Bitcoin

Bitcoin was made-up by an unknown person or group of people with the name of Satoshi Takemoto and release as open-source software in 2009.

Bitcoin is a reward given to miners for mining. The idea of Bitcoin was to have a means of exchange, autonomous of any central authority that could be transferred electronically in a protected, authentic and unassailable way. The domain name "bitcoin.org" was registered on 18 August, 2018. The first Bitcoin was mined with the first block of the chain by Nakamoto in January 2009. The first person who received mined Bitcoin was Cypherpunk Hal Finney who received 10 bitcoins from Nakamoto. According to reports Nakamoto mined an estimated value of 1 million Bitcoin before his vanishing in 2010 when he handed the network alert key and control of the code repository over to Gavin Andresen.

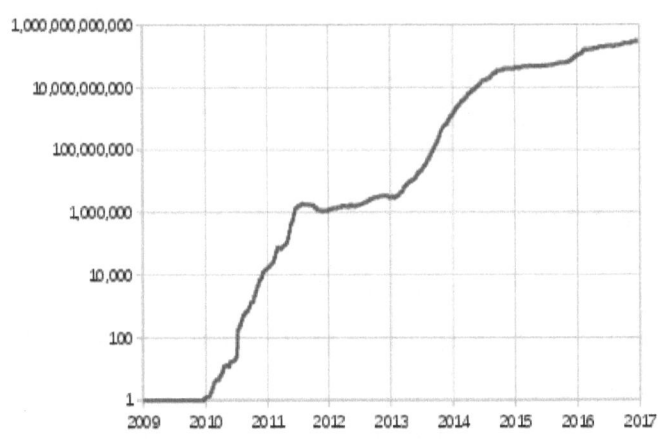

CHAPTER 3

BITCOIN AND TRADITIONAL CURRENCY?

Bitcoin is used in buying and selling provided both parties agree to trade with Bitcoin. Below are some of the difference between bitcoin and traditional currencies:

1. **Immutability:** transactions done using Bitcoin are impossible to reverse,

because there is no central umpire than can authorize the return of the money. An hour after a successful transaction using in the Bitcoin network, it's impossible to modify such transaction. And this makes the bitcoin network a unique means of transacting money.

2. **Decentralization:** one of the unique characteristic of Bitcoin is that, it is decentralized; since no single organization controls the Bitcoin network. The Bitcoin network is being run by an open network of dedicated computers across the world while and maintained by a group of volunteer coders. Persons who are not comfortable with the control of the bank, financial and government institution over their money use Bitcoin since it's decentralized.

3. **Limited supply:** traditional currencies like the dollar, pounds, euro etc. do not

have limited supply. The central bank of such countries can send out unlimited copies of the currencies they want and can tend to manipulate the value of a currency in relative to another currency at the detriment of the citizens who don't have an alternative.

But the supply of Bitcoin is closely controlled by the underlying algorithm. Until a maximum of 21 million is achieved, a small number of new bitcoins will always be mined every hour. This theory makes Bitcoin unique because it works with demand, grows and the supply of Bitcoin remains the same, it will make the value for Bitcoin increase.

4. **Pseudonymity:** when giving money using traditional currencies, payments are usually branded for authentication purposes. But in Bitcoin, you don't need to identify before making any transaction since there is no central

validate. When a transaction application is submitted, the system checks all preceding transaction to be sure you have the available balance to proceed with such transaction.

5. **Divisibility:** the little unit of a Bitcoin is known as Satoshi. And it is one hundred, millions of a Bitcoin (0.00000001). Bitcoin can also be used to enable the micro transactions which the traditional electronic money cannot.

CHAPTER 4

HOW BITCOIN WORKS?

Blockchain:

The Bitcoin blockchain is open ledger that records Bitcoin transactions. The preservation of the blockchain is done by a network of communicating nodes running Bitcoin software's. The Bitcoin blockchains is enforced as a chain of blocks, each block containing a hash of the previous block up to the genesis block. Network nodes can also verify the transactions, add them to their existing copy of the ledger and then transmit these ledger additions to other nodes.

After 10 minutes a new group accepted transaction called a block is created, and included to the blockchain and quickly published to all the nodes. A blockchain is the only place where bitcoins can ever exist.

Transactions:

Transactions which are made up of one or more inputs and outputs and when a user sends bitcoins, the user includes each address and the amount of Bitcoin being sent to that address in an output. Since transaction can have numerous outputs, some users can also send bitcoins to many recipients in a single transaction. Any inputs Santoshis not accounted for in the transaction become the transaction fee.

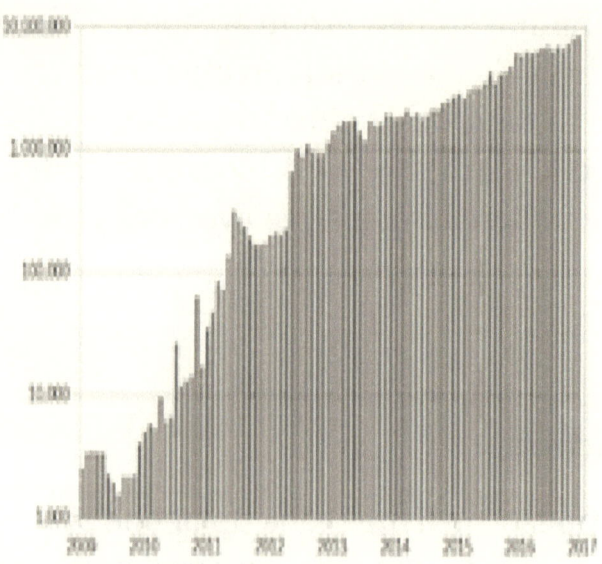

Bitcoin units:

Bitcoin are represented in BTC and XBT symbols. A satoshis is the smallest amount with Bitcoin represents 0.00000001 bitcoins, hundred million of a Bitcoin. A multi Bitcoin equals 0.001 bitcoins, on thousand of a Bitcoin.

Transaction fees:

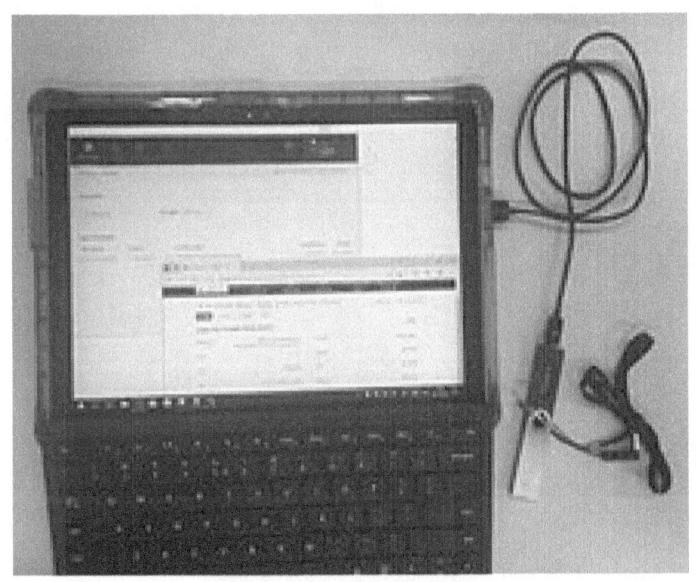

Miners can choose their transactions based on the fee paid and storage size and not the total amount of money paid as a fee. Transaction fees are optional but miners can choose which transaction to go on with and prioritize those that pay higher fees.

These transaction fees are measured majorly in Satoshis per byte (sat/b).

CHAPTER 5

LAWS FOR CREATING BITCOIN ACCOUNT

Bitcoin is registered to Bitcoin wallet address in the block chain. Creating a Bitcoin wallet address can easily be done within few minutes. But to be able to spend your bitcoin in your wallet the owner must know the equivalent key and digitally sign these transactions. The network signs the signature using the public key.

Bitcoin mining:

Mining of bitcoin is a record keeping service done by using computer processing power. Miners keep the block chain constant, total and unchangeable by continually grouping every broadcast transaction into a block which is then broadcast to the network and verified by beneficiary nodes.

The system that is used in mining is based on Adam Back's 1997 anti-spam scheme, "Hash cash". The proof-of-work (POW) requires miners to find an amount called a nonce, so that when the block content is hashed along with the nonce, the result is considerably smaller than the network's difficulty target. The POW along with the chaining of blocks makes amending the block chain very hard, as an attacker must change all the previous blocks in order for the modification of one

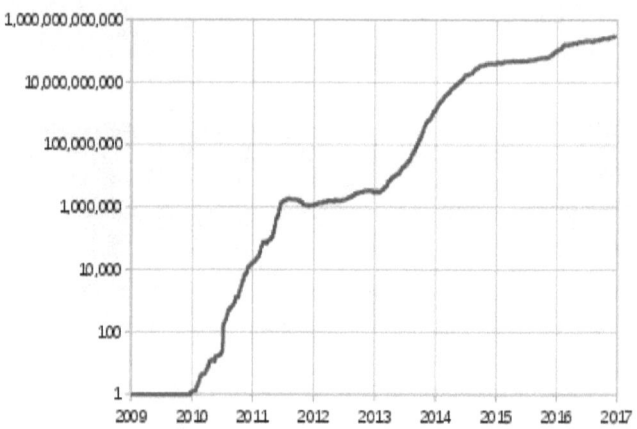

block to be acknowledged. And as new blocks are mined all the time, the difficulty in the process of modifying a block increases as time passes as the number of subsequent blocks increases.

Supply of bitcoin

All successful miners who find the new block are rewarded with newly created bitcoin and transaction fees. To claim the reward, a special transaction called coin base is included with the processed payments.

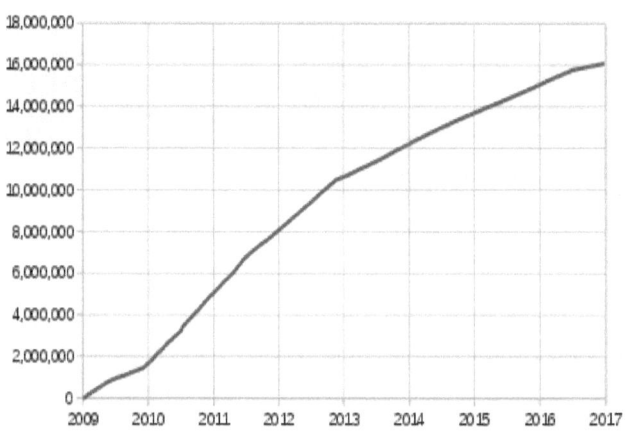

The bitcoin network was designed such that the reward given to miners for adding a block will be halved approximately every four years. As times goes on the reward will decrease to

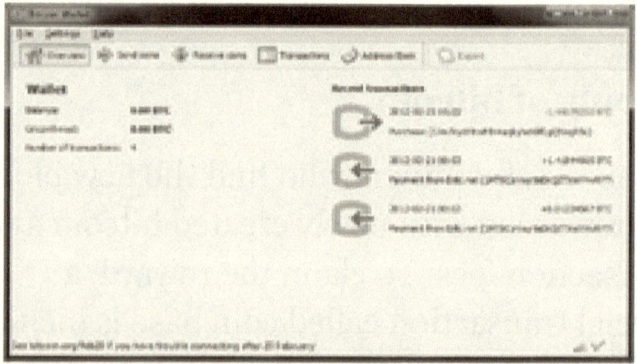

zero and the limit of 21 million bitcoins will be reached, then the miners will be rewarded by transaction fee only. The inventors of bitcoin design a maximum limit of 21 million bitcoins to be ever mined thus generating a synthetic scarcity in the nearest future leading to more increase in bitcoin value.

Bitcoin Wallet:

A bitcoin wallet is like a bank which stores bitcoins, due to the nature of the system bitcoins is always together with the block chain transaction ledger. A bitcoin wallet saves the digital testimonial for your bitcoin holdings and allows one to receive and send bit coi n.

CHAPTER 6

HOW TO PURCHASE BITCOIN?

Now that you have learned the primary things concerning bitcoin and you feel the promptitude to buy bitcoins. Bitcoin can be bought through exchanges or straight from other persons; you can also get them from debits cards to wire transfers or even with other crypto currencies. We shall discuss some steps you can get your bitcoin in the subsequent chapter..

Setup a wallet account

The wallet account allows you to save your bitcoin after purchase notwithstanding your mode of purchase. It can as well be an online wallet that can act as part of an exchange platform or through an independent provider. It can be a desktop wallet, a mobile wallet or offline wallet like a hardware device or a paper wallet).

The most vital part of a wallet is safeguarding and keeping your keys and passwords safe. Losing your password or key may risk you losing access to the bitcoin saved there.

Open an account at an exchange and buy bitcoin online

The exchange company can buy and sell bitcoin on your behalf but only with your permission. There are over hundreds of bitcoin exchanges presently working globally, hence it's very imperative for one to make a thorough research about the exchange company before investing your money, examples of Bitcoin Exchanges Companies known are Coin base, Bit stamp, and Polonies.

Most exchanges company will request for discovery for account setup due to the clampdown on known-Your-Client (KYC) and anti-money-laundering (AML) regulation.

Some exchanges accept payments made through bank transfer or credit card while few accept PayPal transfers. Upon receiving your payment by the exchange company, they will acquire the equivalent amount of bitcoin on your behalf and deposit them in an automatically generated wallet on the exchange.

Purchasing bitcoin with cash

Bitcoins can also be bought with cash using some known platforms. Platforms like LocalBicoins will assist you to find buyers and seller of bitcoins within your vicinity who may want to exchange bitcoin for cash. In the United States platforms like LibertyX will assist in listing out retails outlets who you can give cash for bitcoin easily. Also, platforms like WallofCoins, Playful, and Bit Quick will direct you to a bank near you that will allow you to make a cash deposit and receive bitcoin in few hours.

Bitcoin ATMs are machines that mainly accept and also send through your bitcoin wallet in exchange for cash. Bitcoin ATMs also function in diverse ways similar to bank ATMs you feed in the bills, hold your wallet's QR code up to a screen and the corresponding amount of bitcoin are stag to your account. Coinatmradar can be used to locate the closest ATM to you.

CHAPTER 7

MORE OF BITCOIN

Due to the decentralized characteristic of bitcoin, and its trading on online exchanges located at different parts of the world, regulation of bitcoin has virtually been difficult. Though, bitcoin use can be criminalized in some countries. The lawful status of bitcoin usage differs from one country to another and it's shifting in many other countries. Some countries who have placed either absolute or implicit ban of bitcoin and other crypto currencies includes: Egypt, Algeria, Morocco, United Arab Emirates, China, Indonesia, Qatar, Taiwan, Oman etc. while countries like the US request to register with their financial institution before trading with bitcoin in some of those countries.

Regulatory warning

According to the US commodity futures trading commission advice about crypto currency in December 2017 advise that the virtual currencies are unsafe because:

1. The exchange is not synchronized or supervised by a government agency
2. Market exploitation
3. The exchange may lack system safeguards and customer security
4. Unprecedented price swings and "flash crashes"
5. Theft and hack
6. Self-dealing via exchange

THE END